Bringing a Hardware Product to Market

NAVIGATING THE WILD RIDE FROM CONCEPT TO MASS PRODUCTION

Elaine Chen

ISBN: 1505380839
ISBN 13: 9781505380835

Table of Contents

Testimonials

"At MIT, we like to think that our engineering students know everything about the technical side of starting a company, and all they need is to learn the business side. The reality is that we teach them how to think creatively and invent new technologies, but they don't always have the time and opportunity to experience the full range of all the work that goes into developing a quality hardware product. Elaine has taken her broad experience and boiled it down to a practical roadmap that can help a new hardware entrepreneur know what to expect and to plan for success. It has lots of wisdom from experience and it brings this to the reader in a very practical and useful manner. An extremely valuable read for the hardware entrepreneur."

Bill Aulet, Managing Director and Senior Lecturer, Martin Trust Center for MIT Entrepreneurship; Author, "Disciplined Entrepreneurship"

"Elaine Chen brings lessons from years of leadership building hardware to help the rest of us avoid the most common pitfalls hardware startups face. Hardware takes longer, is less flexible to develop and mistakes are difficult to correct. Learning from Elaine's experience, readers will take away practical techniques that she's used to successfully build wireless controllers, quantified self wearables, and industrial robots."

Eric Paley, Founder and Managing Partner, Founder Collective

"Most first-time hardware entrepreneurs view product development as two steps: prototype then manufacture. Elaine does a superb job of breaking down this false pretense and exposing the complex, multifaceted approach to building great hardware products with a small team. There aren't many books out there that provide practical help for hardware startups; this one is surely a must-read for any aspiring hardware founder."

Ben Einstein, Managing Partner, Bolt

"Elaine draws on years of experience to provide a clear and concise handbook for companies developing consumer electronic products. In this playbook, she helps companies avoid common pitfalls and navigate the challenges of bringing new hardware products to life. It's a must read for both first time hardware entrepreneurs as well as seasoned veterans who want to fill in gaps in their experience."

Scott Miller, Founder and CEO, Dragon Innovation; Partner, Bolt

"Most hardware entrepreneurs have experience in software. And software entrepreneurs have been spoiled. With agile, lean, and modern platforms - software moves fast. Zuck's moto of *move fast & break things* works with software. Hardware is a completely different beast. Breaking things equals expensive recalls! And customers have high expectations for product quality - even from first generation products. Elaine cogently lays out how to build hardware. This is a guide that anyone transitioning from software to hardware should keep close!"

Ben Rubin, Founder and CEO, Change Collective

"This is my go-to reference for all my MIT hardware startups. It takes you through the entire process, in excellent detail, of how to do hardware the right way. I highly recommend it!"

Christina Chase, Entrepreneur in Residence, Martin Trust Center for MIT Entrepreneurship

"Elaine Chen has managed to address the greatest challenge facing hardware startups in a concise manner. Generating great ideas can be accomplished handily during any brainstorming session. The path to product realization however, is akin to climbing Mt. Everest, and doing so before you freeze and run out of provisions. Elaine's step by step guide provides the right amount of theory, practical advice, and doses of reality to accelerate your product to market, and to do so while meeting quality and cost goals. Got an idea for the next big thing? Go ahead and jump off the entrepreneurial cliff, but be sure to read this guide so you can dodge the obstacles and safely land in success."

Adam Shayevitz, President, Strategic Sourcing Dynamics

Preface

In January 2014, I started working with several student startup teams at MIT with hardware product ideas. The interactions often started with this question: "We have a working prototype and are ready to go to mass production, so can you introduce us to a manufacturing partner?"

At this point, the teams usually did a show and tell of the working prototype that they were planning to mass produce. The working prototype often looked something like this.

Figure 1. A duct tape prototype

It broke my heart every time, but I would have to gently tell them that while their technology appeared brilliant and their team was A++, they were missing a few steps from here to mass production. I often ended up drawing the following picture on the whiteboard to depict those missing steps. They had a ways to go.

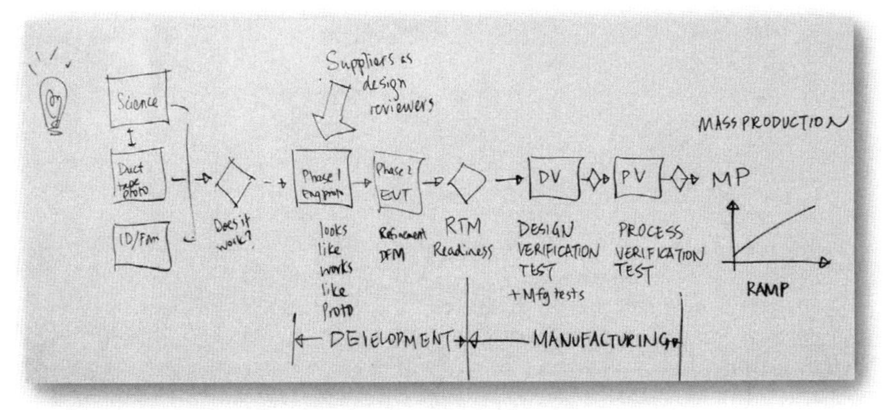

Figure 2. Hardware product development process

For the teams, this was sobering news. Now apprised of how much work remains, they realized they had to replan things from a timeline and budget standpoint.

With this new framework, their journey would still be exciting, but a bit longer and more complex than they had anticipated.

By the time this happened with the third or fourth team, I was frustrated with the lack of practical information to help new hardware product teams get going. Engineering universities do a great job teaching students how to become hardware inventors. But we don't want to talk about the tedious efforts required to take the invention from idea to mass production. Contrast that with software inventors, who can easily find all sorts of help on how to develop and deploy anything on the Web or on mobile apps at any scale. It's not fair.

Developing hardware products is hard enough and costly enough without the added challenge of figuring out how to move forward. So I wrote a blog post, which led to this book.

I wrote this specifically for folks with an engineering background who are experiencing hardware product development and going into production for the first time. I hope it will help product teams understand the process a bit better, and prepare them to gracefully deal with the curveballs that will come their way.

Acknowledgements

I would like to express my deepest gratitude to Alan Frohman, Chuck Brunner, Adam Shayevitz and Michael Caine for patiently reading through early drafts and providing me with invaluable technical feedback to help make sure the material is factually accurate and well balanced. I also want to thank Axel Bichara, Jen McCabe and Bob Steingart for helping me validate my early hypotheses about the needs and wants of the target audience.

Special thanks to my former colleagues at Zeemote, with whom I developed the hand-held wireless controller I used so much to illustrate the work products that come out of each phase of engineering and manufacturing development.

Last but not least, I want to thank the leadership, staff and students at the Martin Trust Center for MIT Entrepreneurship for the incredible privilege of meeting and working with phenomenal student startup teams. Special thanks go to Bill Aulet, the managing director of the Trust Center, for getting me involved in the first place, and Christina Chase, the rock star Entrepreneur In Residence who tirelessly advocates for student entrepreneurs at MIT and elsewhere, and who has mentored countless startup teams.

CHAPTER 1

Why Hardware Takes So Much Longer Than Software

People with a software background who come into hardware development for the first time are often in for a shock. If they practice Agile software development, they are used to seeing significant milestones appear at the end of each one- to two-week sprint. For Web applications, deployment to end users happens continuously, multiple times a day. Mobile applications take a little longer (especially for iOS applications, which need to go through an approval process in the App Store), but each product-release cycle is still measured in weeks, not months or years.

Contrast that, for example, with the development of the Tesla Model S. Although the core technologies had been pioneered in its predecessor, the high-end Tesla Roadster, four years passed between the announcement of the Model S (June 2008) and the first delivery to US customers (June 2012).

The Tesla Model S may be an extreme example. Typical consumer electronics products (such as a wearable device to track activities and sleep, like the Fitbit) can usually be moved from concept to mass production in a matter of six to nine months, if the core technologies are proven. That is still an order of magnitude longer than a typical software product starting at a similar state of technology maturity.

Why does hardware development take so long? Let us look at the process from ideation to mass production.

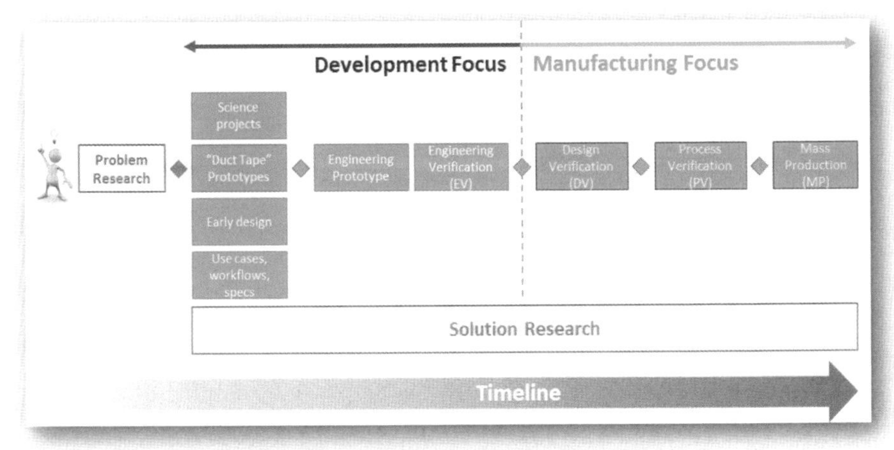

Figure 3. Hardware product development process

This looks just like the waterfall software development model of the bad old days. Unfortunately, there is no good way to adapt Agile software development techniques to hardware development, both from a cadence standpoint and from a work breakdown standpoint.

Here are some characteristics of hardware development that make the entire process long and costly.

- "Big bang" nature of part and assembly design.
  - A physical product with multiple parts that fit together requires each part to be designed in context.
  - No part can be procured and tested until it and all the parts it interfaces with are finalized. Since eventually every part in a design touches some other part, all of the parts have to be released at once, at the end of the design period.
  - Creating and testing part prototypes along the way provides only partial benefit, as each individual part could be fine, yet the entire assembly might not work when it's put together.
  - Contrast that with software development, where a good design makes it possible for functionality to be broken down, developed and fully tested in small chunks.

- Long lead times in procuring custom parts. Some examples:
  - Rapid prototyping (RP) techniques can yield a part in one to three days – less if you own a 3D printer – but the fit, finish, and material properties of RP parts may not be compatible with a functional prototype.
  - Simple machined parts can take one to two weeks.
  - Complex machined parts can take four to six weeks.
  - Sheet metal parts can take two to four weeks.
  - Cable harness prototypes can take two to four weeks.
  - A printed circuit board assembly (PCBA) can be prototyped in short production runs in one to three weeks.
  - Production lead times for everything are categorically an order of magnitude longer than the above.
  - Contrast that with the time to prototype a new software feature – which can range from a few minutes to a few days.
- Interdependencies that cannot be parallelized.
  - Each prototyping phase is complete when the entire product is assembled, integrated, tested, and debugged, and when it yields a functional prototype.
  - The functional prototype is required for effective engineering validation and user testing.
  - Results from engineering validation and user testing are required to define design changes for the next iteration.
  - Consequently these components in the overall program can be hard to run in parallel, making it hard to compress the overall schedule.
- System integration is always longer than expected.
  - For hardware products with custom software, unexpected side effects always come up when each prototype is assembled and tested for the first time.
  - None of these side effects show up when modules are tested in isolation. Here, again, component level testing and validation are only partially effective – the big test remains the final integration and debugging process.
- Prototyping and tooling costs.
  - Prototypes can cost a lot of money. Short-run builds are never cheap.
  - When it's time to go to production, developers must invest in tooling to take advantage of mass production processes like injection

molding or die casting. Tooling can cost from $500,000 to a few million dollars, depending on the number and scale of custom parts.

- Hurdles before getting to a saleable minimum viable product (MVP).
  - Hardware products must meet baseline quality, durability, safety and regulatory compliance requirements before they can be sold legally, especially in Europe.
  - Crowdfunding has made it possible for product teams to sell prototypes directly to backers, and these prototypes bear much less scrutiny than actual saleable products. It is a great way to validate early market interest, but it does not remove the need to go the last mile in quality and regulatory compliance.

All these factors contribute to making hardware development slower and more expensive than software development. It behooves any hardware product development team to really understand the phases of the process and how they relate to each other. That way, teams can make the best decision at each step of the way, and do everything they can to parallelize development, maximize knowledge of the market, and shorten the time from concept to mass production.

We will look at each of these phases in detail in the next few chapters.

CHAPTER 2

On Primary Market Research

You may have noticed the two prominent boxes that have to do with research in the flow chart below.

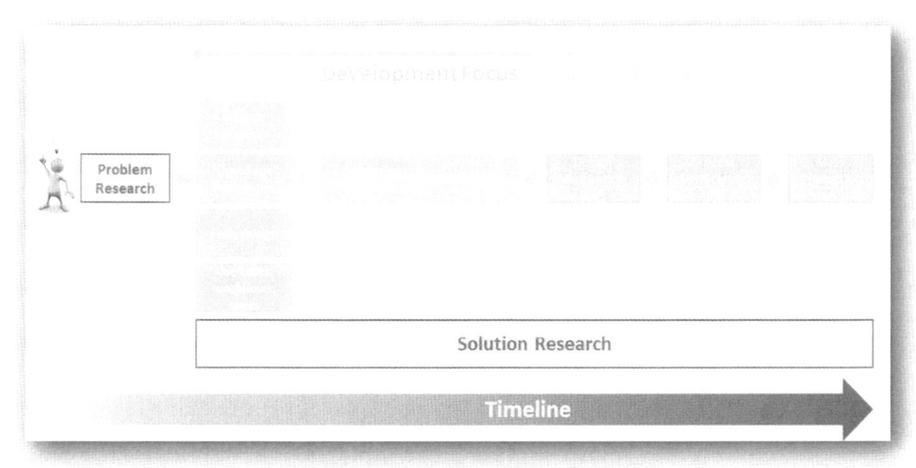

Figure 4. Problem and Solution Research

Here's a question: Why is "Problem Research" happening before early technology development?

At the beginning of product development, the most important question is not the "how," but the "who," "what," and "why." Focusing on *how* to build the product and getting a good head start on implementation without a thorough understanding of the market and the customers is a strategy that runs a high risk of producing a beautiful product that nobody wants or needs.

Instead, the product team should focus on *who* the customers are, *what* their needs, wants and expectations are, and *why* your product solves their problems.

The only way to understand all that is to get out of the building and do some primary market research. Let's look at the types of research that can help a product team define the right solution for the target market.

There are two kinds of primary market research in product development: problem research and solution research.

In problem research, you begin with the objective of understanding the buyer and user personas, what their needs, wants and expectations are surrounding your area of interest, and how they make purchase decisions. The product is typically not shown during this process. You are trying to understand the people and the market, not trying to sell the product.

Solution research is conducted after the idea has taken shape but before any significant product development efforts have started. It is OK to do solution research in parallel with scientific research and advanced engineering development, but you must be absolutely sure to validate all your hypotheses before beginning the final stage of the engineering work.

It is healthy to continue to do problem research throughout the program – especially during the early stages of product development. However, the intensity of this activity usually tapers off once the development efforts gather momentum.

In solution research, you show mockups, prototypes and early product builds to potential users and buyers to get feedback on functionality, usability, and pricing. The product is shown, but you are still not yet trying to sell the product – you are trying to gather feedback to define and tweak the product before you lock and load on creating it.

Solution research should happen throughout the development cycle. Test early and often, with any stimulus materials you can get your hands on. (Stimulus materials are artifacts generated for the research project or produced as a byproduct of normal development efforts.) While solution research should be ongoing, time should be devoted to focused testing whenever a major integrated prototype comes available.

Qualitative research techniques

The most effective primary market research employs qualitative research techniques.

One common technique involves conducting one-on-one detailed interviews with potential end users of the product or service, and with prospective buyers, as well as anybody who has significant influence over the buying process. The research team would first come up with goals and objectives for the project, write up a protocol, develop a discussion guide, and figure out who will visit whom and who will be the chief interviewer. In the actual interview, the research team sits down with its research subjects one at a time, asks open-ended questions, and watches and listens attentively to the responses.

Founders should do their own research

Often, technical founders with an engineering background feel intensely uncomfortable about going out to the market and doing qualitative research themselves. Their education and training prepared them to deal with numbers and facts, and talking about feelings and emotions with someone they don't know can be quite disconcerting.

Sometimes this discomfort causes the founders to outsource the whole research project to professionals (e.g., a primary research consultancy), and they never actually get out to see customers themselves. This is a mistake. Good product insights come from deep understanding of the market and customer, and there is no way to get that level of understanding with second-hand research results.

However uncomfortable you may feel, you should still attend at least some, if not all, of the early customer interviews. Get someone else to be the lead interviewer if you must, and become the note taker, but do go and meet these people first-hand. That's the only way to really understand the problem you want your product to solve and therefore create a true solution.

A few tenets

- Do primary market research early and often, and do a lot of it.
    - Due to the high costs and long lead times to arrive at a saleable hardware product, it is paramount that development teams embrace a learning stance and engage in customer research continuously throughout the development cycle.
- Do not confuse problem research with solution research.
    - Keep in mind that problem research is about the problem, and solution research is about the solution. Problem research comes first. Once you start talking to someone about the product, you have said goodbye to any chance of a deep conversation about needs and wants – the conversation will be fixated on the product concept instead.
- It is not a sales call.
    - You are not there to convince people that your idea is fantastic – you are there to learn and listen. Talk very little and keep your eyes and ears wide open.

There are many primary market research techniques that can be employed for each of these two types of research. I will discuss them in later chapters.

CHAPTER 3

The Development Phases

So you have conducted your primary market research, and you have a good idea of the target market, user personas, and their needs, wants, and expectations. You've come up with a few ideas on how to address these needs and wants. You are now ready to go into development mode, starting with research and advanced development.

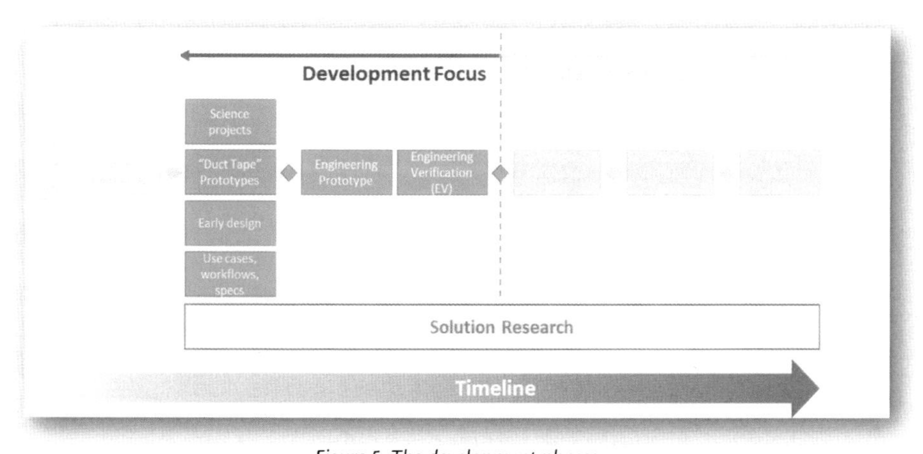

Figure 5. The development phases

For this discussion, let's assume we are working on a hypothetical consumer electronics device, to be sold in the US. Let's assume the product is a battery-powered connected device, involving custom sensors and custom printed circuit boards inside custom plastic housings. It has a cost of goods below $50 US, and you hope to produce a volume of 10,000 to 50,000 units in the first 12 months.

Let's further assume the product will be assembled, tested, and shipped by a third-party contract manufacturer with whom you have developed a partnership. The location of the contract manufacturer (US or offshore) doesn't matter in this discussion.

For a product like this, there are generally three conceptual development phases.

Phase 0: Discovery and feasibility

In this phase, the product team works together to reduce technical and market risk and to come up with enough information to help define the minimum viable product (MVP), which will be developed in the subsequent engineering and manufacturing phases.

Developing a duct tape prototype

For most hardware products with significant technical risk, the technical team starts by constructing a series of quick engineering "breadboards," also known as duct tape prototypes.

The objective of these prototypes is to prove the efficacy of the science and technology behind the idea. Prototypes generated at this phase work somewhat like the final product, but often look nothing like it. Technical viability and basic considerations of human factors and ergonomics are the focus. The full user experience comes later, after the technical risk has been adequately addressed and the team believes there is a viable path forward.

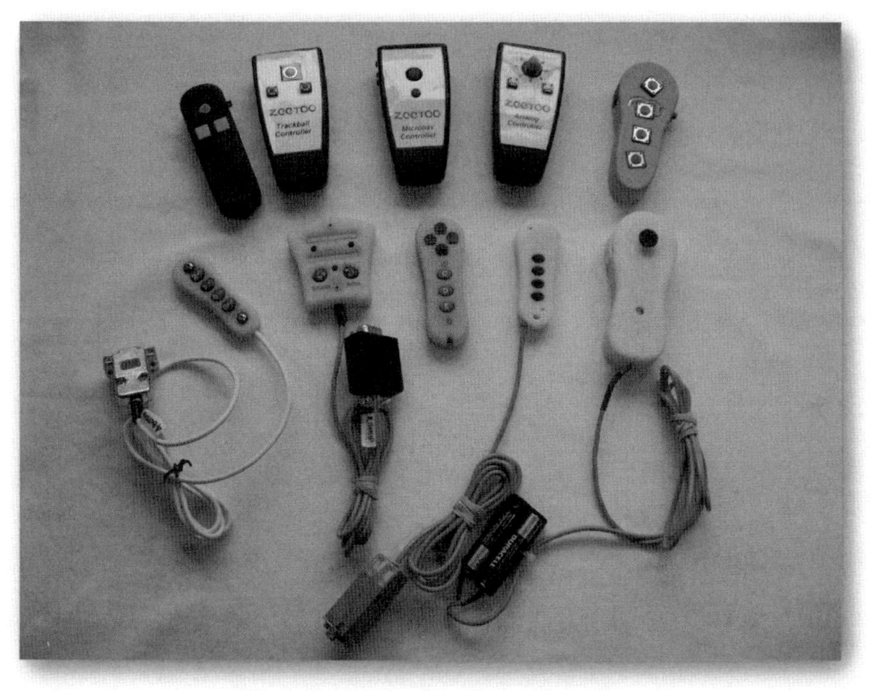

Figure 6. Examples of duct tape prototypes

Since firmware development requires a custom embedded platform, schedule overruns may occur when unexpected delays on the hardware platform spill over into the development time for integration and firmware.

One way to avoid such delays is to use development kits from companies selling a key component—for example, a development kit with a Coretex-M3 chip in the same family as your final selection, or a Bluetooth development kit.

Figure 7. An evaluation board for the STM32F103MCU from STMicroelectronics, a leading microprocessor manufacturer

These development kits are usually much larger than a custom board with the same component in the final design. They are loaded with peripherals you don't need, and are not appropriate for creating a fully integrated prototype with electronics inside custom housing.

However, they can help the firmware team get a head start in learning the development environment for that chip without developing a custom board, waiting for the board to be fabricated, and then having to integrate and debug when the final specifications of that custom board are still in flux.

Vaporware

In parallel with the duct tape prototyping phase, the team should also be working on early stage user experience and form exploration for industrial design, to help drive system architecture decisions for the next stage.

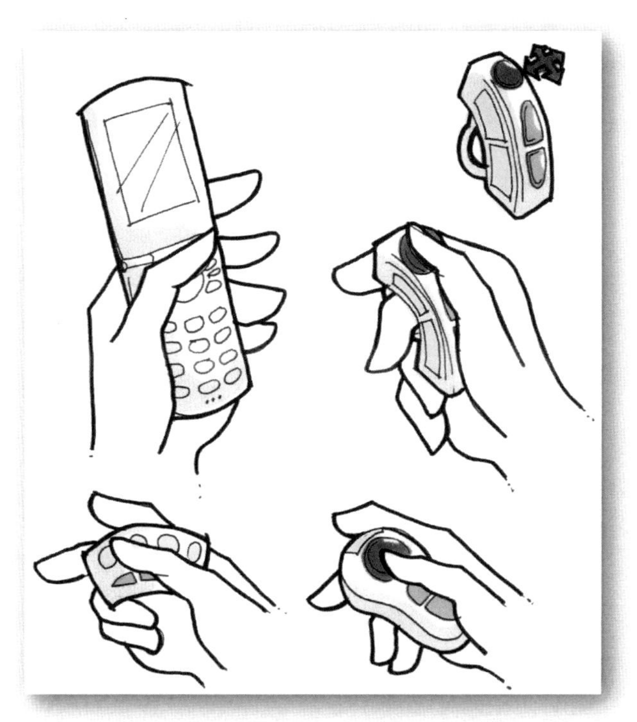

Figure 8. Examples of marker renderings capturing the ideation process

Sometimes the form exploration and early industrial design generate foamcore, foam, or printed physical models. There is always value in getting right at the physical form, especially for hand-held products that must be sensitive to human factors and ergonomics. However, the form should not get ahead of the engineering architecture.

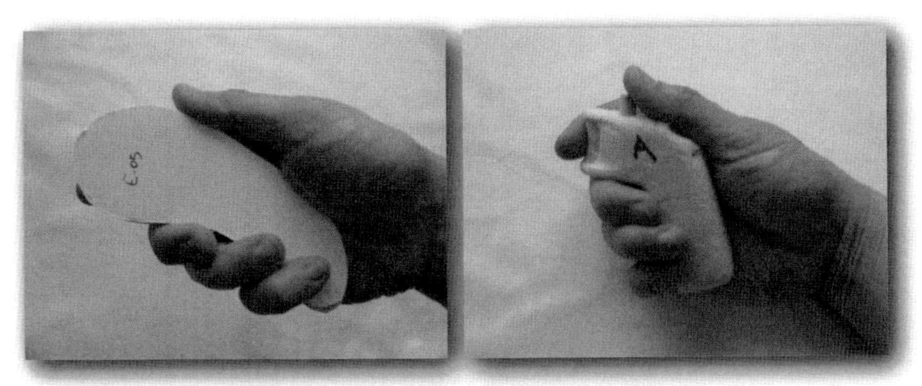

Figure 9. Early foamcore and foam models of a hand-held product

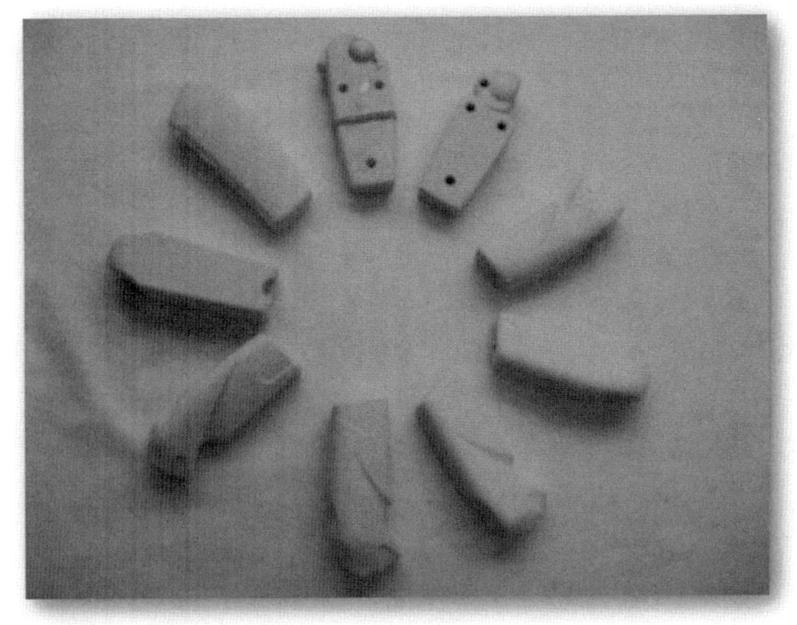

Figure 10. 3D printing and other rapid prototyping techniques can help the product team quickly create and test large numbers of forms with small variations

A good time to start playing with form is after the duct tape prototyping phase yields acceptable results and the system architecture of the product is defined for the interior technical package in rough volumetric forms. This gives the industrial design process a jump start and helps to compress the elapsed time to program completion.

Ongoing primary market research

Primary market research should be strongly featured in this and all future phases – both problem research (to determine who the customers are and what their needs and wants are) and solution research (to determine whether the proposed solution is usable, useful, and saleable).

Basis for Development gate review

There should be an explicit gate review at the end of Phase 0 to decide whether to move on to the next engineering phase. This is often called the "Basis for Development" gate.

Three criteria are necessary to move forward:

- First, technical feasibility must be proven. You don't want a science project in the middle of a development schedule.
- Second, market research must support the business case for moving forward with this product.
- Third, a requirements document or functional specification for the minimum viable product must be developed, articulating the solution that the team believes will answer the customer problems identified through the research process.

A word on product requirements

"Requirements" has become something of a dirty word in product-land since Agile software came into vogue. It has become associated with antiquated, slow-moving processes. But there is a very good reason to invest in a brief requirements document prior to starting a big engineering effort: It is one of several tools for building effective cross-functional leadership.

Most modern-day hardware development is massively cross-functional. With any program of moderate scale, you can easily be coordinating a team of five to fifteen people at any given time, not to mention external partners like the contract manufacturer, the suppliers, and the labs that test your product for regulatory compliance.

It is critical that everyone has a clear understanding of the systems-level end goal before they proceed with their individual components.

A brief product requirements document created at this stage, and updated continuously in the next two development stages, to reflect new understanding of the market, is the fastest, cheapest and easiest way to put down some stakes and keep everybody in sync about the big picture.

I am not suggesting you write a 300-page document – but a 3- or 4- or even 10-page document would be a very good investment to set you and your team up for the next stage.

Remember that the big engineering spend starts in Phase 1. A bit of planning goes a very long way to make sure that the money is spent efficiently and that the development time is used effectively.

Phase 1: Engineering Prototype

At the start of Phase 1, all the key questions should already be answered: market receptiveness, technical feasibility, key system architecture decisions and a full understanding of the product requirements necessary to create a winning solution. The goal of Phase 1 is to take the product definition from Phase 0 and bring it to life in a fully integrated, looks-like, works-like prototype, with a design intended to be production-ready. This is usually the most exciting phase for a design engineer, who can finally see the product come together for the first time.

This phase involves a coordinated effort for industrial design, mechanical, embedded engineering (printed circuit board, cable harness development, firmware development) and software engineering to work together in a cross-functional program plan.

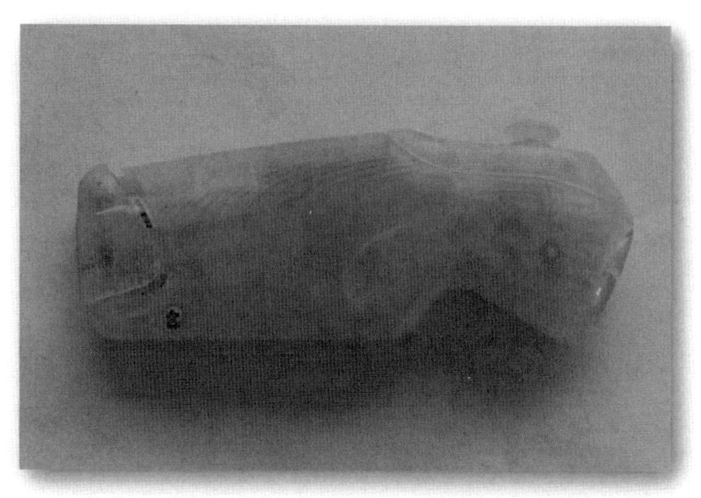

Figure 11. A typical example of a Phase 1 engineering prototype. The fully assembled prototype represents production intent part design. Custom parts are typically produced using rapid prototyping techniques.

Leveraging rapid prototyping techniques to create custom parts

Rapid prototyping (RP) processes are usually used to create custom parts for this phase because they deliver a fast turnaround.

- For plastics, rapid prototyping may involve 3D printed parts, stereolithography, or plastic laser sintering.
- For metals, it may involve machining, even if the target manufacturing process involves casting or extruding the final part.
- For printed circuit boards there are many quick-turn fabricators that can create a small-run surface-mount technology board for a fee.
- Cable harnesses are usually produced using prototyping techniques as well. (What? There aren't any cable harnesses? You are hand-soldering wires? Stop and hire an embedded engineer with production experience immediately!)

A side note on using RP processes to create custom plastic parts: While RP is fantastic for fast turnaround, as cited by Eric Ries in his book, *The Lean Startup*, as a viable way to support the build-measure-learn process, custom parts made via RP processes are usually not appropriate for use in a final product, for several reasons:

- The cost for any given plastic part made with an RP process can be 20 to 100 times more than the cost of that same part made using mass production techniques, such as injection molding.
- The material properties and achievable tolerances for RP parts are not always appropriate for the final design, particularly if the part is used for structural purposes. For instance, the stereolithography process produces very brittle parts that degrade over time. Fused deposition modeling processes that create thermoplastics parts have better material strength but have much coarser tolerances. As of 2014, I am not aware of any RP process that can create a glass-filled or graphite-filled plastic part for material strength purposes.
- As of 2014, the finest resolution 3D printing, stereolithography, or plastic laser sintering process on the market still produces a surface finish that is inferior to injection molding – and consumers have come to expect the high quality fit and finish of injection molding.
- Even if you can find a magical RP process that meets all your requirements – cost, material properties and achievable tolerance – the speed with which each part can be made is inadequate for high volume manufacturing. For instance, it takes less than 60 seconds to create one or more plastic parts from a single cycle on an injection-molding machine, but it can take 30 minutes to print the same part using an RP process.

Having said all that, the world of RP and 3D printing is dynamic, and one day additive manufacturing may become a viable manufacturing process. In fact, some industries, including aerospace, are already adopting 3D printing of metal parts for some of their manufacturing processes, creating parts that are compatible with the quality of output and the speed of part creation.

The procurement process for prototyping will look completely different if and when 3D printers can print with the high quality fit and finish and structural strength, at an acceptable speed and with reasonable costs.

Final assembly

Final assembly is typically done in house by the engineering staff. No one hands that task to outside manufacturers or even in-house electromechanical technicians in the engineering lab.

Why? First, this is a brand new design, and the only people who have any hope of putting it together successfully are those who designed the product.

Second, nothing comes together correctly the first time around, and tinkering is usually needed to get the assembly to fit at all. Tinkering, in fact, informs the design engineer about any changes necessary to make the next iteration fit correctly.

Third, assembling a design for the first time is self-actualization for any design engineer. You would not be able to pry the engineer away from the parts as they arrive.

Last but not least, the assembly process is an important way to learn about design-for-manufacturing matters. You want only members of your team to be the ones gaining this knowledge.

Prototype count

Typically, people create one to five copies of the prototype at this stage. It is a first-generation prototype and there are bound to be issues: design errors, fit/finish problems, or outright functionality problems. It does not pay to make 50 copies when you might have to trash 49 of them.

Design for manufacturing and cost reduction considerations

Sometimes design engineers mistakenly believe that it is possible to design a product quickly just to get the design intent fully captured in a working prototype. They expect to then do a round of design tweaks to render it production-ready and to reduce the cost of goods sold (COGS) for the product.

Don't fall prey to this thought process. If you have a target for your final COGS, perhaps because of immutable pricing constraints from a business standpoint, you need to design it right into the first prototype. Likewise, the quickest route to a producible design for a quality product is to incorporate design-for-manufacturing (DFM) thinking into the engineering.

The best DFM strategy I know is to design every custom part with the target manufacturing technique in mind. Then, start interviewing suppliers for each part, and engage top candidates in design reviews early and often.

Getting suppliers involved from the get-go ensures that the parts that are designed will be compatible with the manufacturing processes and the capabilities of the manufacturing partner. It is the single most effective way to minimize the distance from Phase 1 to Phase 2, accelerating time to market and reducing the cost of non-recurring engineering. It also starts building a relationship between your company and the supplier, which will be a key player in your future operations.

Getting a head start on strategic sourcing

Although this is only an engineering development phase and not "real manufacturing," it is still a good idea to work with a supply chain expert to source all key components from original equipment manufacturer (OEM) suppliers now, instead of three phases later.

You may be able to get the job done faster using, say, Digikey as your primary supplier, but that will entail a massive redesign when you need to bring down the cost, and you will end up having to switch out some components. Or, for example, you may have fallen in love with a design requiring a 19-inch touch screen display, only to discover that the demand for 19-inch displays is waning and the cost is prohibitive. In that case you would have to change the entire industrial and mechanical design to accommodate a smaller screen.

Note that strategic sourcing is not the same as supply chain management, which is not purchasing. Also quite different from each other are manufacturing,

manufacturing engineering, manufacturing test engineering, quality control, and logistics. A lot of engineers fail to understand that "manufacturing" is a massive discipline with multiple specialties. Mixing up these functions would be akin to mixing up what a mechanical engineer and a software engineer do on a day-to-day basis. Product folks with no manufacturing background should read up on the topic to get a basic sense of these disciplines. This will facilitate better collaboration and cross-functional communication throughout the program.

The need for speed: toward which milestone?

In my opinion, if the ultimate objective is to minimize time to market for a saleable hardware product, it is far cheaper and less painful in the long term to use the engineering Phase 1 to get your cost down, incorporate DFM principles, and tentatively settle on your strategic sourcing.

Of course, your mileage will vary by the program you are running and the primary objectives of your engineering prototyping phase. There are times when it makes sense to put aside DFM and cost considerations and instead prioritize speed of implementation.

One such scenario is when you are trying to create a concept prototype to quickly showcase your vision in order to get buy-in, funding or approval for the other stages, which could take months or years to complete. In that event, going for speed and a high quality prototype would make sense. Just be careful of expectation management – if you show a prototype that looks great and works adequately during a demo, and your fundamental design direction results in a COGS that is 10 times off plan, you will have some challenges during the next phases.

Dealing with DFM, cost reduction and strategic sourcing early makes sense only if the program is definitely a go, and you are trying to minimize the time to market. My suggestion is to use your judgment and make these decisions based on what you are trying to optimize. Maximizing the speed to a first working prototype does not usually involve the same steps as minimizing the elapsed time and cost and maximizing the quality of output of an entire product development program.

System integration

Because this book has a hardware focus, I will not belabor the software side of the equation. However, I do need to mention integration as a specific challenge unique in developing connected devices (or, for that matter, any hardware product with a significant amount of integrated software delivered in an embedded fashion).

With any hardware prototyping effort, the first challenges are to get all the parts to fit together and to assemble everything so it stays assembled. The result, however, is but a beautiful sculpture until the firmware is finished and the full system has been integrated.

For a connected device with a cloud component, the integration also means verifying that any Web service application programming interfaces or APIs work over wire, or wirelessly with a WiFi connection. The integration process usually takes several weeks to a couple of months, depending on the complexity of the system. It requires many planets to be aligned: the firmware, software, electronics and mechanical components all need to be individually developed and tested as much as possible prior to the actual system integration.

Every product team I have worked with underestimates this phase. Don't. This is the single most important and strategic activity in Phase 1 and it can determine the functionality and usability of your product,. Skimping on this phase will only add time to the end of the schedule.

For the connected device example, a good rule of thumb is to expect system integration to take at least one-third to one-half of the time it took to do your hardware design, procurement, debugging and assembly, especially if you are testing a new product platform for the first time. This ensures that you will understand holistically the characteristics of what you have built.

Leveraging solution research to support design iterations

The engineering prototype phase is the best time for the product team to employ the build-measure-learn methodology in the Lean Startup framework to help evolve a minimum viable product.

While the product team would have developed requirements with the best information available, the work is largely based on gut, speculation, and extrapolation. Solution research with various artifacts generated during this phase will help to tighten the product definition and ensure that the solution created is useful and usable to end customers.

However, the process of creating a fully functioning prototype is often not divisible into smaller chunks that can be tested independently. For instance, the user experience may be testable only after system integration, which could be weeks or months into this engineering phase. The opportunity to test in the market could be reduced to a tiny window in between engineering phases.

With huge pressure to minimize time to market, a lot of product teams end up adopting an execution mindset and shut out primary market research activities altogether. These teams may suffer the consequences of prematurely and irreversibly committing to a design that isn't right for the end customer.

Balancing the need to minimize time to market and the need to build the right product is always challenging. Here are some ways to incorporate as much customer feedback into the process as possible without waiting until the very end. The trick is to test the product concept with customers every step of the way – well before the final prototype is assembled.

- Test hardware forms with storyboards and sketches via an interactive interview process, then iterate as needed.
- Create form models that don't have the requisite functionality, and test for human factors and ergonomics using rapid prototyped parts with no internal mechanisms. Iterate.
- Test software usability with interactive mockups that do not have a fully functioning back end. Iterate.
- Test the entire user experience with components created in Phase 0. Modify product definition if necessary.

The above methods allow components of the system to benefit from the build-measure-learn loop rapidly without having to wait for the entire cycle of procurement, assembly, integration, and debugging to conclude. That said, the ultimate test can be done only with the final prototype.

I would suggest that product teams set aside two to four weeks between Phase 1 and Phase 2 to rigorously test the fully integrated prototype. The teams should be open-minded and balanced in interpreting the feedback.

In the best scenario, the program allows for multiple engineering prototype iterations within Phase 1. The product team can get key lessons from component testing, and the team can do a full design iteration prior to entering the engineering verification phase. However, the pressure to minimize time to market sometimes makes multiple full iterations impractical. Product teams will need to do the best they can with the time and resources available to make sure they are strongly tuned in to the market and customer needs.

Design review

While there is not usually a formal executive gate-level review at the conclusion of this phase, it is wise to have an internal design review with the product team at the conclusion of Phase 1.

The review should include a demo of the integrated prototype, running an early version of the production software.

Any findings from the assembly process, as well as solution research in the field and in the lab (e.g., usability research), should be integrated to help generate a short list of things to tweak for Phase 2.

Confronting inconvenient truths

Ideally, all the customer research up to this point has resulted in a product that is on the right track, and any lessons learned won't require massive architectural changes. But if the product testing, either via primary market research or internal engineering validation, raises significant issues about the chosen approach, it behooves the product team to take a deep breath and seriously contemplate a replan.

The ultimate measure of success for any product team is not whether the product hits the planned ship date, but whether the product will succeed in the market. It is far better to take a schedule hit to adjust to your findings about the market than to blindly adhere to a first-customer ship date and end up building the wrong product.

Phase 2: Engineering Verification (EV)

The EV build is a design iteration of the engineering prototype that addresses a targeted short list of issues raised during the assembly, system integration and testing of the first-generation engineering prototype.

The EV build is usually identical to the final saleable product for all practical purposes. The form, function, surface finish, colors and textures should all represent the final production intent design.

Figure 12. The EV build looks like a saleable product, although it is not saleable at this stage

Minimize the scope of change from Phase 1 to Phase 2

The best possible Phase 1/Phase 2 transition involves a tightly controlled scope of change between the two designs. Ideally, the EV build looks very much like the Phase 1 design.

This is because the larger the scope of planned changes, the larger the probability of new issues created by those changes. In the worst case you could end up inserting a Phase 2.5 to fix those issues (prolonging time to market and burning up more of your cash runway than you can afford – resulting in the need to raise more funds to bridge the gap).

Can you combine Phases 1 and 2?

Phase 1 in this framework is the first time that someone tackles the detailed industrial, mechanical, and embedded systems design of the planned product, and Phase 2 is an incremental iteration to catch mistakes. One might ask: Can you just get all of that done with a carefully executed Phase 1?

It really depends on the complexity of the product and the amount of invention and innovation that the new design represents. A product that involves well-understood components that have a well-known manufacturing process (e.g., a wireless mouse) probably does not need a full-on EV phase. But a product that has a significant amount of invention, and particularly a product that has complex electromechanical components, stringent environmental requirements, and lots of moving parts (e.g., the Scooba, a floor-scrubbing robot from iRobot), will definitely need a Phase 2 Engineering Verification phase to address design issues identified in Phase 1.

Manufacturing techniques for custom parts

Prototyping techniques continue to be appropriate in Phase 2. At higher volumes (more than 10 copies), crossover low-volume manufacturing techniques become very appropriate. One such example is urethane casting – the use of a machined or stereolithography "pattern" to create silicone or rubber molds, which are then used to create castings of a plastic part at a lower cost than rapid prototyping, which makes the part one at a time.

Final assembly

Final assembly is still done in house. This is because there are design changes from Phase 1 to Phase 2, and it is essential for the engineering staff to check the integrity of the design and tweak anything necessary to ensure that the product can be manufactured and works as intended.

Typical prototype count at this stage

The EV stage is an engineering phase, so the prototypes still tend to be expensive. Assembly is also expensive because it is done in-house by engineering staff. So it is usually cost prohibitive to produce very many. Ten to twenty is a good target.

Engineering verification tests

The industry acronym for the EV phase is actually EVT – engineering verification test – for a very good reason: this is the production intent build and represents the first opportunity to fully exercise the functionality of the product and conduct both engineering and durability tests to validate the integrity of the design.

Taking our battery-powered connected device idea a bit further, let's say the device is a wearable bracelet with a fabric strap and has a plastic housing for an electronics module. It also has a custom sensor for some biometric signal (e.g., heart rate) and runs with a cloud back end and a mobile front end. The hardware/systems part of the tests might include:

- Sensor characterization
    - Quantify resolution of the sensor.
    - Quantify absolute accuracy and repeatability (if applicable).
    - Quantify the linearity of the sensor readings and mitigate if necessary.
    - Quantify environmental effects, such as radio frequency, temperature changes, etc.
- Buttons
    - Test all buttons and knobs to ensure they work to specifications.
    - Design a life testing system to push the button x million times, or whatever capacity your design calls for, and test to failure.
- Radio
    - Test and validate range of the Bluetooth radio at various battery levels.

  - Test all modes of operation, including pairing, unpairing, connecting, disconnecting and any other modes necessary for normal operation on all supported smartphone platforms.
  - Test edge cases – what happens if the connected device and the mobile phone get out of range? What if the mobile phone runs out of battery?
- Battery
  - Quantify the time and output voltage/current levels of the battery as it drains down.
  - Test the system's performance at various battery charge levels and mitigate if necessary.
- Environmental
  - If your product has an Ingress Protection or IP rating, send it out to test against IP standards for that rating. For example, if the product is supposed to be watertight, test to see if it stays watertight.
  - Test the product to the full range of its specifications relative to temperature range, humidity and other environmental conditions.
- Fabric testing
  - Dye-transfer test – ensure chosen fabric does not bleed.
  - Stretch test – ensure stitched fabric strap is elastic enough and won't come apart.
  - Durability test – stretch fabric and relax it by the number of rated cycles.

Basis for Production gate review

At the end of Phase 2, there should be a formal gate review to demonstrate the EV build, review findings from field tests and early durability tests, double-check findings from ongoing market research, and make a go/no-go decision on releasing the design to manufacturing. This is sometimes called the "Basis for Production" gate.

This is the most critical gate review in the entire development project.

It is the crossover point where the program shifts from an engineering focus to a manufacturing/production focus. Ownership of the program shifts from the engineering department to the manufacturing and operations department after this gate.

If the product involves tooling, this is the point at which the team decides whether the design is ready to be released to tooling – triggering a large capital investment in custom tools that take a long elapsed time to create.

One way to visualize the significance of this gate review is to look at the overall program costs. From ideation to this gate review, you may have spent one quarter of the total program cost. You are about to trigger the remaining three quarters and this is pretty much an irreversible decision. Given the investment and elapsed time at stake, it makes sense to give this gate review a lot of thought.

Confronting more inconvenient truths late in the game

Sometimes, despite best efforts every step of the way, new data emerge during the EV phase that give the product team pause. Say there is mounting evidence from the market that suggests the product as implemented misses the mark on a key parameter. There may be new results from engineering verification tests indicating there is a fundamental flaw in the technical architecture.

How an organization responds to such inconvenient truths can have a huge effect on the ultimate success of the venture. At this late stage, the cost of pulling the plug would appear prohibitive. The venture could well exhaust its funds or "run out of runway" and need additional capital to keep going. There would be every temptation to stay the course and keep the original first customer shipment date.

However, if significant new data challenge fundamental assumptions about the product and the business, it is the fiduciary responsibility of the product team to raise the issue, carefully consider the implications and, if necessary, hit the pause button and take as much time as necessary to contemplate next steps.

In some cases, suspending the project may not be possible because of funding requirements or business conditions. Then you have to keep moving forward, incorporate what changes you can absorb and leave the rest for the next product iteration.

Nevertheless, stakeholders need to be fully informed of the facts and the likely consequences for the success of the product in the market.

Starting the manufacturing process is starting to scale the venture. You want to make every effort to scale with the right product in hand.

CHAPTER 4

The Manufacturing Phases

Once the engineering design has been validated and finalized and the team has decided to move forward based on the best data available, you are ready to move into manufacturing.

Primary ownership of the program passes from engineering to manufacturing at this point. The supply chain, including the contract manufacturer (CM), has been established and will become very active in the program now.

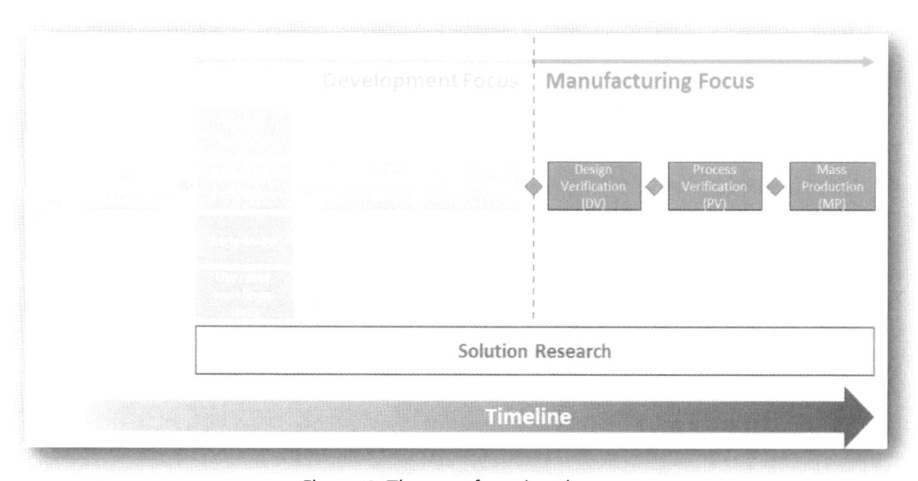

Figure 13. The manufacturing phases

There are generally three manufacturing phases in a mass-produced product:

Manufacturing Phase 1: Design Verification (DV)

The primary purpose of the DV phase is to teach the contract manufacturer how to build the product and, in so doing, identify any remaining design-for-manufacturing issues. There should be few differences from EV to DV.

The parts used in the DV build should be virtually identical to the EV build. The only difference is that the EV build is typically assembled by in-house staff, whereas the DV build is assembled by the contract manufacturer's staff.

Manufacturing techniques for custom parts

Prototyping techniques are still appropriate for this phase, because the tooling cycle takes months and fully tooled parts are typically not all ready when the DV build needs to be assembled. That said, some early parts might come in time for this build.

Final assembly

The contract manufacturer will need significant on-site support from the engineering staff to build the prototypes.

Typical prototype count at this stage

By this time the business functions are clamoring for more copies of the prototype so they can use it for testing or at trade shows and other venues. The engineering function wants more copies for life testing, durability testing, destructive testing, emissions testing, and drop testing. Myriad types of engineering validations will continue in advance of the start of mass production.

There will be tremendous pressure to make a lot of copies at the DV stage. Don't. I have seen companies go up to 50 copies at this stage, but they paid for it later. I still prefer 10 to 20 copies for this phase.

Why limit copies?

- The primary purpose of the DV phase, once again, is not to create copies of prototypes, but to teach the contract manufacturer how to make the product and devise an efficient process.
- Making more than a handful of copies will distract the contract manufacturer and the engineering team from this learning process, prolonging the time to market and possibly lowering the quality of the first few lots of product coming off the manufacturing line.
- The labor costs and opportunity costs are lower than the EV build, but the part cost is unchanged, because the majority of tools are in development and you are still paying through the nose to get custom parts made via RP techniques.

Culture shock

If you come out of an engineering background, as I did, you may find yourself in unfamiliar territory when you work with manufacturing professionals on your first product.

Most design engineers are driven to quickly iterate and improve their design, and they find it very hard not to fix something that they can see a way to improve. They may not fully appreciate the processes and cross-dependencies downstream, and thus become frustrated when their need for speed meets apparent resistance.

However, the manufacturing process is predicated on the premise that the design is frozen and well documented, to facilitate excellent execution of a quality production process. Any design changes, no matter how small, cause significant challenges for the manufacturer and must be carefully controlled through an engineering change order process. Each change request triggers a series of events on the contract manufacturer's side of the house, largely opaque to the product team, that may result in IT systems being updated, visual instructions or standard operating procedures being rewritten, and manufacturing technicians being retrained.

As a result, even extremely simple tweaks, such as changing the type of fastener in a subassembly, can take two to three weeks from start to finish on the manufacturing side.

This can cause significant tension between the engineering and manufacturing organizations. Engineering thinks manufacturing is slow-moving and bureaucratic. Manufacturing thinks engineering is going cowboy and endangering the program by making unnecessary changes (or worse, sneaking in changes without adequate documentation).

To get to the finish line, all of these perspectives need to be accommodated holistically. The trick is to work hard at understanding the constraints and concerns under which the other team operates and limit the changes while still finding a way to expedite critical fixes.

Tool development / First Article Inspection

In parallel with teaching the contract manufacturer how to assemble the product, the engineering team must also support tooling development for component suppliers in charge of creating custom parts. This is where the multi-month tooling cycle starts.

Figure 14. Cavity side of a tool for a custom housing part. This is the side that creates the surface that faces out.

It is incredibly gratifying to visit a supplier and see the actual tools being created, and exciting too to prepare in-house for first article inspection. While this can require a bit of back and forth, with tool tweaking and the need for documentation, this is the home stretch. After the tools are all set, the custom parts are ready to be mass-produced.

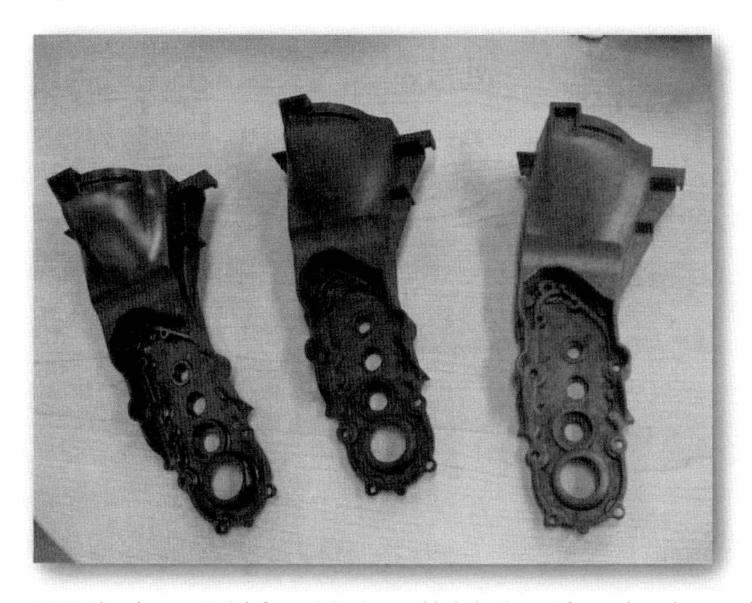

Figure 15. Testing three materials for an injection-molded plastic part from a brand new mold

Process development and manufacturing test development

The CM now needs to work with your team on process development, determining how to make your product with a high quality of output. It is paramount that the design stabilize at this time. Any changes should be minor and must be tightly controlled and documented via an engineering change order process.

The DV phase is also the time to develop assembly fixtures, as well as any test fixtures and manufacturing test and calibration software to be used during the build process.

Figure 16. A classic "bed of nails" automatic PCBA tester used in the manufacturing process

Supporting regulatory testing for product marks

DV prototypes are often appropriate for use in regulatory testing. For instance, if your product plugs into a wall outlet, you will need a safety mark from a US government-approved Nationally Recognized Testing Laboratory, which will test your product to the appropriate safety standards.

If your product has a radio inside, you will need emissions testing to meet Federal Communications Commission requirements and to get an FCC number, which must be displayed on the product. If you are selling your product in Europe, it must have a European Economic Area CE mark. There are exceptions to these rules, but they are generally non-negotiable for consumer electronics products sold on sites like Amazon or in stores like Best Buy and Apple.

You can test for compliance only when the product is fully finished, but it is always a good idea to test your product for emissions and robustness of electrostatic discharge events as you go – as early as the breadboard phase. This will help you avoid having to revamp embedded systems at the eleventh hour.

Figure 17. Symbols that indicate a product has undergone compliance testing and regulatory approval

DV to PV Readiness Gate Review

At the conclusion of the DV phase, there is typically a gate review for process verification (PV) readiness.

The engineering design, manufacturing test systems and manufacturing processes are examined for readiness to move on to the next production phase.

Manufacturing Phase 2: Process Verification

The PV build tests out the manufacturing processes devised in the course of the EV and DV phases. Ideally, there should be no engineering design changes from EV to DV, all the way to PV and MP. It's all about process from here on.

That said, new product development is all about invention, and typically the PV gate review finds issues that need to be addressed. These need to be carefully evaluated case by case. Critical fixes that affect the product's quality, yield and manufacturability should be allowed, while discretionary design changes should be held for a next iteration of the product.

The PV build will involve parts made via the final manufacturing process, in order to ensure that the end-stage process will create high-quality parts with good fit and finish to support a saleable product.

Depending on how the preceding phases went, as well as the lead times for custom tooling, this phase could be as short as a few weeks or as long as a few months.

Packaging

One thing that happens during this phase is the packaging design, an often overlooked but very important detail. If your product will be displayed in a physical retail store, the design must be appropriate for placement in the store – e.g., a product package that hangs on a peg will need to have a flat-form factor. If your product needs to be drop-shipped, its packaging is the shock absorber that protects the product during shipping. You will do well to contract out the packaging engineering to a professional service provider to make sure the packaging serves its purpose.

Figure 18 Packaging example suitable for a retail store display

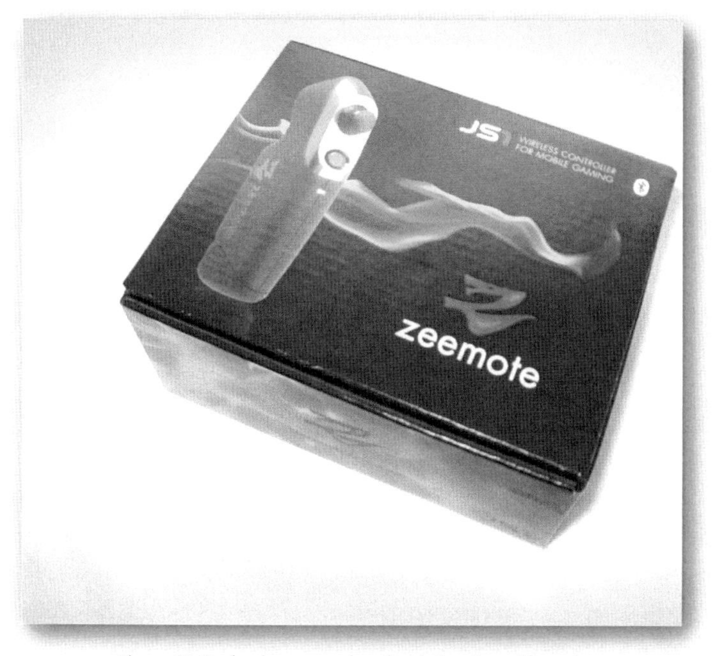

Figure 19. Packaging example suitable for drop ship only

"Basis for Market" gate review, which doubles as a mass production gate

Once the process verification test is complete, you and your CM must decide whether to move toward mass production. This is the final gate review.

This is also the time when your marketing department gets ready to launch the product, although sometimes the decision to go forward with marketing launch can precede the mass-production gate by a few weeks.

Deciding to proceed with the marketing launch and naming a shipping date can be risky. Any production delays can disappoint customers who have been primed to expect the product at a certain date, and whose expectations need to be reset. It is usually best to withhold the marketing launch until there is very good clarity on the probable first customer shipment date, and to hedge on the exact date if there are any doubts.

Manufacturing Phase 3: Mass Production (MP)

If you get here, congratulations! You made it!

The next steps are as follows:

- Energetically find and fix all the "infant mortality" issues in the design and manufacturing process. These are issues that cause the product to fail, right out of the box. You might see a rush of them during the PV phase, but they would typically calm down in the months after the start of mass production.
- Continue to test finished goods at a high sampling rate, and find and fix new issues over the next six to nine months. This sustained activity is jointly owned by engineering and manufacturing, but shifts to manufacturing over time, as design changes for the product stabilize.
- Launch an ongoing reliability test program where you sample finished goods over time and put them into durability test fixtures. The early units provide valuable systems-level reliability data for the product as assembled in the production process. The later units provide a longitudinal view on product quality and reliability over time.
- Once there is clear evidence that quality issues are largely addressed, start to scale up the capacity of the production line to support the volume ramp required by the unit forecast.

Low-incidence defects

You are likely to have to spend six to nine months fixing a long list of latent design issues or process problems.

This surprising fact may sound depressing, but it's simply a numbers game. You will have found and fixed a lot of "infant mortality" issues in both the design and the prototyping phases. But even after that, low-incidence defects will provide considerable work for both engineering and manufacturing for months.

After all, if there is a defect that occurs at a rate of 1 in 1,000, and you make only 200 of your product a week at the outset, it may take months to properly characterize the defect and come up with a mitigation measure.

This is to be expected. Every hardware manufacturer goes through it, even Apple. (Remember Antennagate?) So plan and budget for it, and don't get annoyed. It's a normal part of the manufacturing ramp-up.

"Rightshoring"

Offshore manufacturing used to be the de-facto choice for any hardware startup aspiring to achieve a healthy gross margin. The cost of goods sold (COGS) for offshore manufacturing always looks attractive. In recent years, however, there has been a lot of dialogue about whether this makes sense for everyone and everything. The short answer is that it doesn't.

Different ways to design and manufacture a hardware product

I have managed the development of products that are manufactured in myriad ways:

- Designed in the US and manufactured in-house by staff assembly technicians.
- Designed in the US and manufactured by a contract manufacturer (CM) in the US.
- Designed in the US and manufactured by a CM in Asia serving as an original equipment manufacturer (OEM) partner.
- Designed and manufactured by a CM in Asia serving as an original design manufacturer (ODM) partner.

Picking the right strategy

In my personal experience there is no one-size-fits-all strategy. However, there are factors that will significantly influence the onshore/offshore manufacturing decision.

The most important factor is the relationship between production volume and the value of the product. Are you creating a low-value, high-volume product such as consumer electronics? Is your per-unit cost of goods under $50 US? Are you ready to commit to making at least 10,000 units in the first year of production? If so, offshore manufacturing is probably the right path. On the other hand, if you are creating a high-value, low-volume product such as a medical device with a hefty COGS (say $1,000 US) and a relatively low volume (say 5,000 a year for the first year of production), you should seriously look into manufacturing in the US.

A second factor is the level of intellectual property protection you need. Are you working on a government contract? If so, you are definitely manufacturing in the US. You may have to choose a CM who is a certified defense contractor, with everyone on staff a US citizen with security clearance. That is an extreme case, but if you have intellectual property you are looking to protect, you may want to avoid offshoring. Case in point: An exact replica of one of my prior products popped up in Asia a few years after we started producing them in southern China. Somebody somewhere leaked the technical package. There was nothing anyone could do about this, because intellectual property laws are largely unenforceable in China. If this is a significant concern, manufacturing in the US will give you more peace of mind.

A third factor is the precision and level of invention needed in process development. Does your product have lots of moving parts? Do you need high-precision parts? If the answer is yes, then process development is a lot more significant. The counterexample is commodity products like laptops and mobile phones. These involve fully validated and low-risk manufacturing processes. Honing a complex production process is far more easily done if the CM is close to the R&D.

What can be made in the US?

As a general rule, military products, which are usually produced in low volumes, need to be made in the US. High-value products such as medical devices and luxury goods -- made at any volume – are comfortably producible in the US. Low-volume products need to be made in the US because the business model of low-run manufacturing is incompatible with the majority of CMs in Asia.

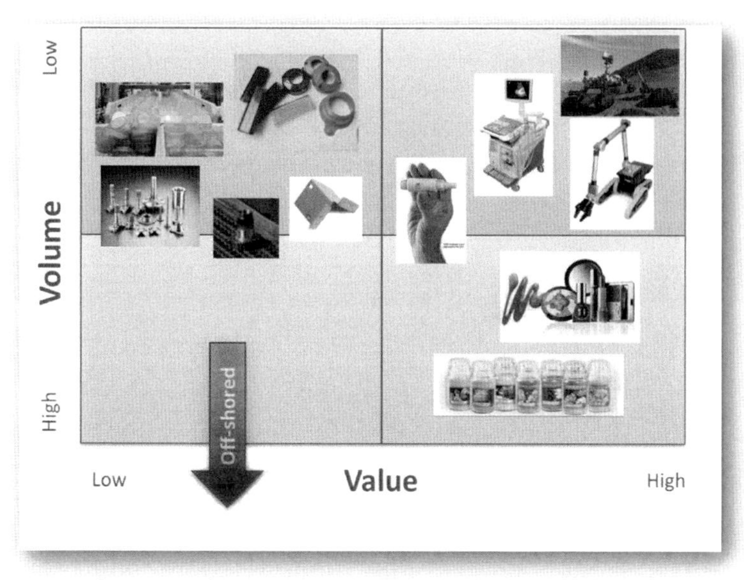

Figure 20. Made in America: Military products; high value products; low volume products

Low-volume, low-value products are the most challenging to manufacture. While making these products in the US makes the most sense, it is hard to get the COGS to an acceptable level. This is because much of the cost is in the setup of the manufacturing line. This is often a fixed cost regardless of the quantity of products. At low volumes, this setup cost gets passed on to the startup in the form of a higher COGS, significantly reducing profit margin.

Not having proof of a viable COGS makes it very hard to develop a sales-based business model. This is an important problem, for which several groups in the US and Asia are trying to find solutions. This issue will be worth revisiting in a couple of years to see if things have improved.

The hidden costs of offshoring

Startup principals may not realize that offshore manufacturing comes with a lot of hidden costs. The expense of repeat trips, with airfare and per diem charges in two- or three-week stays for multiple team members, adds up quickly. The lost-opportunity cost of having team members frequently travelling is huge. Different time zones can mean longer turnaround times for resolving critical issues. Having $500,000 in inventory tied up for six weeks while it is being shipped to the US can present a significant cash-flow management challenge. It is critical that you take the time to calculate the total cost of ownership of your manufacturing initiative; COGS and tooling costs are only part of the equation.

Another issue with startups looking to manufacture their first products in Asia is a business model mismatch with their CM. The startup is looking to move very fast: producing in small lots, fixing issues as they arise, and iterating as manufacturing proceeds. The CM, however, values stability of the product and works best with large runs of a mature, stable product.

Due to this mismatch, CMs often find that they lose money on startup deals. Startups typically lack a credible credit record and are not good at making their production forecasts. Their volumes are lower and their products are less proven, resulting in a greater risk of low-yield or catastrophic stop-ship events in the first year of production.

The CM has a much easier time dealing with mature companies and products. When push comes to shove, the CM often ends up prioritizing its big customers over its startup customers and may become non-responsive to the needs of the startups. While this tendency exists for all CMs, it seems to arise more in Asia, where large-scale manufacturing is more common.

In the worst-case scenario, the CM could decide that it is losing money and proceed to "fire" the startup from its customer base. Since most startups are single-sourced to their supply base, this could be a huge problem. This has happened to several startups I know. It's not pretty.

Transitioning from the US to offshore manufacturing

If you have a new product that you are looking to test in the market, and you have low volumes to begin with, it is better to start manufacturing in the US. Keeping manufacturing close to the R&D team is invaluable. Once the market and the product have stabilized, you can reassess your decision.

An interesting way to tackle the transition to manufacturing offshore is to work with a multinational manufacturer. Such companies have a presence in the US and multiple locations offshore. This allows you to start manufacturing in the US at a higher COGS level and – staying with the same contract manufacturer – transition to an offshore facility if and when your volumes and business model support this move. That way, you can have flexibility while you are developing a new product, and a cost-competitive COGS when you are producing at scale.

CHAPTER 5

Organizing The Troops

This is a good time to discuss cross-functional leadership in product development. Let's use the same product example discussed earlier in this book:

> *For this discussion, let's assume we are working on a hypothetical consumer electronics device, to be sold in the US. Let's assume the product is a battery-powered connected device, involving custom sensors and custom printed circuit boards inside custom plastic housings. It has a cost of goods below $50 US and you hope to produce a volume of 10,000 to 50,000 units in the first 12 months.*

The goal is to staff the program for success and maximize the performance of the project team.

In this example, we have a highly cross-functional program with multiple objectives, which require team members to come from different functional disciplines. You will need people with specific expertise for each objective. For example:

- To gather market insight and understand the customer:
  - Qualitative researchers
- To develop the hardware and build prototypes:
  - Industrial designers
  - Mechanical engineers
  - Electrical engineers
  - Firmware developers
  - Electromechanical technicians

- To develop the packaging:
    - Graphic designer
    - Packaging engineers
- To develop the software and associated algorithms:
    - Web and mobile software developers
    - Information architects
    - Interactive designers
    - Graphic designers
    - Data scientists
    - Database administrators
    - Software quality assurance engineers
- To produce the product and manage the operations:
    - Manufacturing engineers
    - Manufacturing test engineers
    - Sourcing professionals
    - Supply chain management professionals
    - Purchasing professionals
    - Logistics professionals
    - Production quality control engineers

In this environment, a silo mindset – where specialists stay in their comfort zones within their own functional discipline – almost guarantees project delays and budget overruns. It is critical that you develop the team structure, culture and dynamic that encourage members to identify initially with the broader team and share their pursuit of the Holy Grail, instead of first identifying with their functional disciplines, focused on their daily tasks at work.

There are a few things to consider when organizing such a diverse group of people.

- If you are the chief technical kahuna of this effort, keep in mind that you don't need to know everything about every discipline. In fact it is not possible. You simply need to find great people, provide a solid framework to coordinate work, and trust each person to pull his or her weight. But you must know how to delegate work and gauge progress
- It is paramount that the team is culturally set up to work fluidly across functional disciplines. For example, the mechanical part designer needs

to have access to, and be comfortable talking to, the software engineer who is writing the code to control the mechanism. The engineering disciplines must be tightly coupled with the manufacturing disciplines from the get-go. This needs to be addressed at the top level; senior managers should articulate their expectations about keeping communication channels open.

- Set up governance structures that foster and enforce cross-functional collaboration. For instance, consider generating a "scrum of scrums" standup meeting for leaders of individual areas, in order to coordinate work at a higher level than the regular scrums with the troops.
- Define goals and success in terms of the overall program, not the individual team's deadlines and deliverables. For instance, the design's release to manufacturing from engineering is one measure of success for engineering, but the larger goal is arriving at mass production with a good quality product. Avoid defining too many small goals that are specific to each functional discipline, as it may cause folks to optimize locally while the program suffers globally.
- Communicate broad program goals frequently to all team members to keep everybody in sync, and always discuss issues in context of their significance to these goals.
- When doing customer research, create an "observer" spot and rotate all teams' members through it, so everybody gets contact with the customer and gets to see the end goal.
- Make it easy for people from different backgrounds to build relationships with one another. Some ideas:
    - Make project teams sit together regardless of discipline. Co-locate designers, product managers and different types of engineers who work on site.
    - Create events where people will hear from somebody they don't work with on a regular basis. For example, organize a lunch-and-learn series and get different people to give a talk each time.
    - Celebrate program milestones by organizing mini-parties for everyone involved. Ice cream... cupcakes... beer... all are good choices.
- Assign a project manager to create, maintain and continually publish an overall schedule with key milestones so that everybody, from the intern

to the principal software engineer, knows where things stand and whether things are happening according to plan. This can be a team member, a dedicated project manager, or the engineering lead (my preference).

- Be mindful and deliberate about the choice of structure for organizing people and projects.
  - First and foremost: regardless of the size of the team, the buck stops somewhere. If one executive is in charge of all disciplines, it stops either at the VP of Engineering or VP of Product Development. If there are two executives, each of whom is in charge of some of the disciplines, it stops at the CEO or COO. It can work perfectly fine either way, although the latter works best when the CEO or COO has a technical background.
  - For a cross-functional project team of five to ten people, keeping them all as one team works great.
  - For ten to thirty people, you will need some substructure.
    - One way is to assign team leads for subsystems, so each subsystem is still cross-functional but the number of people working closely together is inside the five to ten range.
    - Another way is to assign team leads by discipline but with the project organized in a matrix format. This is a little more difficult to do because the functional organization will pull people back into their comfort zone.
  - For more than 30 people, you have to have sub-teams for the broader team. The same options apply.
    - Sub-teams or tigers can be created to tackle different subsystems, if this is feasible. For instance, a robot team can have a gripper sub-team, an arm sub-team and a base sub-team.
    - Organizing troops by functional discipline and managing the project as one initiative is the more classic way to organize. For instance, the robot team can be subdivided into the mechanical team, electrical team, controls team, software team, and so forth.
    - If more than one project is going on simultaneously, it may make sense to separate into project teams, giving each person a primary project affiliation. This is often done when there are R&D

or advanced development activities that need to be managed at a different cadence and with different techniques.

        - One last way to organize work is the matrix organization, most commonly found in product design consultancies. Everybody has one line manager and a different project manager, and each person can be working on multiple projects at once. Use caution. One has to be trained to work and manage in a matrix setting, and it doesn't work at all when there is too much work for too few people. Not my favorite.
    - For 50-plus: to keep the agility and dynamics of the small team structure, I prefer dividing up the people by project or business unit. That is the cleanest way to make the priorities clear for everybody. It keeps the team cross-functional, and it limits multitasking.
- You may not need everyone to be a full-time employee. To stay lean, you can outsource large chunks of the work to experienced contractors. However, I would suggest you identify your core competencies and key differentiators, and keep those functions in house.

CHAPTER 6

A Word On Customer Development

Lean Startup/Customer Development is a great framework for a market- and customer- driven approach to entrepreneurship. As many people know, this approach was introduced to the technical startup world by Eric Ries, author of *The Lean Startup*, and Steve Blank, author of The *Four Steps to the Epiphany* and *The Startup Owner's Manual.*

The basic tenet of customer development is to invest heavily in customer research and underinvest in product development. With this strategy, you rely on a low-fidelity minimum viable product to test the business model, and iterate until you get it right before launching into full-fledged product development.

There are four steps to this process:

- **Customer discovery** – Work to understand who the customer is and search for problem/solution fit.
- **Customer validation** – Start selling the minimum viable product and iterate until a repeatable sales process is found.
- **Customer creation** – Grow customer base from early adopters into the mainstream.
- **Company building** – View your organization structure and build out the team to execute the business model.

The Customer Development framework further stipulates that these four steps involve the search for a business model using a build-measure-learn philosophy. After a scalable business model is arrived at, and product/market fit is found, execution follows. Execution is defined as actual product development, involving product management and planned engineering development.

This works all the way to Step 2 for hardware development, and then we run into issues.

Let's discuss these four steps again with respect to a hardware product, using my connected-device consumer electronics product as an example.

Step 1: Customer Discovery.
Any product person focused on success would start here. Too many startups become preoccupied with the technology or their concept of the product and miss the boat on understanding the market and the customer's needs and wants. "Stop building the product and get out of the building, now!" is what we say to startup teams over and over.

Step 2: Customer Validation.
For a product with hardware components, this is where we run into trouble. In the phrase "repeatable sales process," the key word is "sales." For a software product that has, say, a cloud back end, some big data algorithms and a mobile and Web front end, you can begin iterating on the sale process immediately with a very minimal, low fidelity-MVP. For a hardware product, *there is no such thing as a low-fidelity MVP*. Would you:

- Buy a brand new eco-friendly dishwasher for your house if you heard that 30 percent of the shipped units leak sudsy water all over the kitchen?
- Buy a new, non-invasive blood glucose meter for your grandparent with diabetes, if it produces a correct reading little more than half the time?
- Buy a new car that parallel parks by a novel mechanism that rotates the wheels 90 degrees and scoots directly into narrow street-parking spaces, even though it lacks crash and durability test results because they were delayed to minimize the time to market?

On top of these considerations, there are the not-so-small challenges of regulatory approval and product marks. If these words are new to you, I invite you to turn over your computer's power supply and have a look at the sticker. You will most likely see a "UL" mark. That is a fire and safety compliance mark that tells customers your product can be safely plugged into the wall without blowing stuff up. If you turn over your mobile phone (you may have to open the battery compartment for Android phones), you will see a number of other product marks, like FCC and CE. The former is an emissions mark necessary for any product with a radio inside (2.4GHz... Bluetooth... Cellular... they aren't picky) to be legally sold in the US. The latter is a compliance mark that allows a product to be legally sold in Europe.

There are nationally recognized laboratories that test for safety, emissions and other applicable standards before issuing these product marks. Without these marks, you cannot sell the product legally. And the regulatory approval process for a new product can take four to twelve weeks.

To repeat, for hardware products, there is no such thing as a low-fidelity MVP. You either make a quality product or forfeit your opportunity to test a real business model. The best you can do is to collect donation funds in return for providing a prototype to potential customers. You cannot list on Amazon, and your product cannot get placed in Best Buy or Apple or Target or Walmart without undergoing testing and approvals.

The other point that gives me pause is the idea that you do not need to hone your execution engine to create a product, which again is not applicable to hardware or software products of any reasonable complexity. Coming back to the connected-device example, the development process is long (measured in months), costly (measured in hundreds of thousands of dollars, frequently exceeding $1 million even for simple products) and fraught with risks. If you do not have well-tuned product management, development, and manufacturing organizations that are excellent at execution, you will find yourself in a quagmire of project delays and budget overruns. You could run out of money before you can ship to your first customer (the dreaded "tail wagging the dog" in the Customer Development framework).

So hardware folks are stuck in the stage gate process. What can we do to embrace key tenets of Lean Startup thinking to maximize our chance of success?

- Truly invest in Step 1 – customer discovery – before embarking on product development. Take as long as you need to truly understand the market and the customers. Look for comparables. Founders should definitely do this themselves, and on a very tight budget. An idea with no market is a non-viable idea. Learn how to do qualitative research properly. Conduct detailed contextual interviews. Do observation studies. Do immersive studies. Do anything and everything to maximize your understanding of the customer before you proceed.
- Embrace the M in MVP when defining your product. Avoid feature bloat: put in the fewest possible hardware features, thus reducing development time and risk.
- Design the product in such a way that it can become more and more capable with a series of software releases running on the same hardware platform.
- Incorporate continuous customer research as an ethos while working on Step 2, customer validation, which starts with creating a saleable product. Test anything and everything with customers. Test with storyboards, foam models, printed models, or duct tape prototypes while waiting for the first engineering prototype to come out. Plan on four to six weeks of customer testing time between the engineering prototype phase and the engineering verification phase. Be prepared to pull the plug and go back to Step 1 if you discover anything about the product that disproves your hypothesis.
- During the first engineering prototyping phase, employ rapid prototyping techniques to iterate through design concepts as quickly as you can, and test every build as you go.
- For a connected device with custom electronics, consider making a semi-Frankenstein prototype using external development kits to circumvent the lead times for custom printed circuit board assembly procurement. For example, you can buy Bluetooth modules that will be too big to fit into your wearable connected device, but you can fake the functionality by running a wire from the module to your wearable devices and get going with customer research that way.

- Drag your feet on building out the rest of your organization beyond product, engineering and manufacturing until relatively close to first customer shipment. Consider using contractors instead of full-time hires for sales and marketing personnel, in case the sales model changes. For instance, channel sales are different from direct sales, and the personnel you would recruit and the organization structure would be vastly different. Hedge if you aren't sure which way will stick.

Having done all that, you could still end up with a product that doesn't fly the way you think it should. Coming up with a second product to test a different business model would be months or years out. So the proper build-measure-learn loop is long and hard to do for a hardware startup. But by incorporating Lean Startup and Customer Development principles throughout the development cycle and making time for customer research, you will increase your odds of success more than if you did everything inside the building.

Summary

In this book, we explored the traditional phases involved in developing a hardware product that is mass-produced using standard manufacturing techniques, as well as some tips and tricks to maximize the probability of success via the continuous application of primary market research.

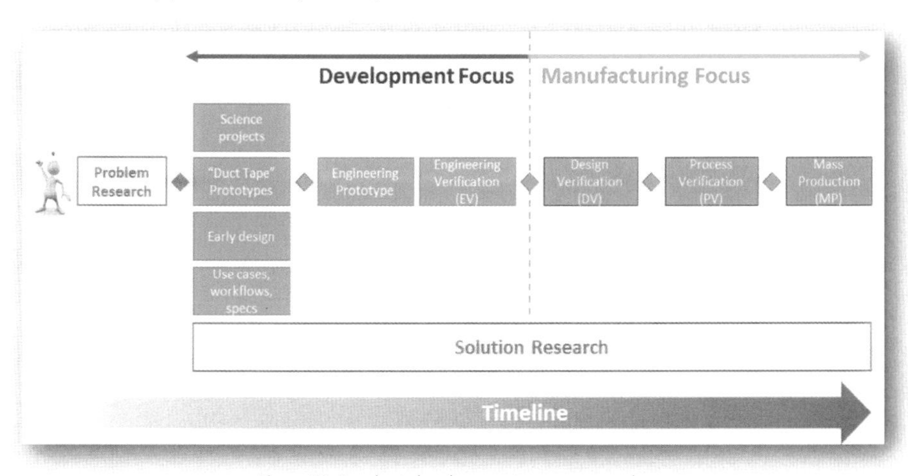

Figure 21. Product development process, reprise

The process can be long and considerably more convoluted than the process for a software product, which can arrive at a saleable minimum viable product in a matter of weeks. It is very rare to see a hardware product reach that milestone in less than three to six months, especially if regulatory approval is involved.

However, it is incredibly rewarding to see a tangible product come out of the factory and into the hands of customers. And you now know all the steps to get from here to there, as well as a few tricks to accelerate the process. Go forth and create some great hardware products!

Figure 22. Zeemote JS1 wireless controller for mobile gaming

Glossary

primary market research	Original and customized market research tailored to a specific industry, market segment, buyer or user for a product or service. For instance, interviewing 50 college students to see if they own an Android or iOS mobile device is primary research.
secondary market research	Collecting and interpreting preexisting market data that already exist. For instance, reading a published report from a research analyst about Android versus iOS market share in a given time period is secondary research.
problem research	A primary market research effort to build knowledge about a market segment, buyer and user personas, and their needs, wants and expectations.
solution research	A primary market research effort to test a specific product or service solution with end users and potential economic buyers, to collect data about pricing, functionality, usability and more.

Lean Startup	A movement started by Eric Ries, author of *The Lean Startup*, in which the build-measure-learn loop is used to test and hone the minimum viable product (MVP) in the market with real customers until the product arrives at product/market fit.
Customer Development	A movement started by Steve Blank, author of *The Four Steps to the Epiphany* and *The Startup Owner's Manual*, emphasizing the importance of getting out of the building and working with customers to build knowledge about the problem and solution and to develop the sales model.
build-measure-learn	A learning framework whereby a product team forms a hypothesis to test, builds an iteration of the minimum viable product, goes out to test this hypothesis, and synthesizes the learning into specifications for the next iteration.
minimum viable product (MVP)	Eric Ries defines it as the smallest possible thing that can be developed to facilitate a complete build-measure-learn loop.
minimum viable business product (MVBP)	In his book *Disciplined Entrepreneurship*, Bill Aulet defines this as the smallest possible product that can be sold to test a business model.
rapid prototyping (RP)	A series of techniques that quickly produces custom parts without a lengthy tooling process (often plastic, but metal parts are increasingly supported by new processes).

stereolithography	A rapid prototyping process for custom plastic parts with very good resolution and tolerances – good for checking the fit for cosmetic parts in a final assembly, but generally not appropriate for structural parts, due to the brittle material properties of the substrate.
selective laser sintering	An RP process for custom plastic parts with great material properties and a rather coarse surface finish – good for structural parts, not good for cosmetic parts. Material is also more prone to warping than an injection molded version.
3D printing	A general term for a rapid prototyping process to produce "grown parts" by depositing beads of melted substrate one layer at a time, building up the part additively.
contract manufacturer (CM)	A manufacturing partner who will mass-produce a product under contract.
engineering breadboard (aka duct tape prototype)	Phase 0 of the development cycle, in which engineering invests in rapid feasibility studies, focusing on technical functionality instead of the full package.
engineering prototype	Phase 1 of the development cycle, in which a looks-like, works-like prototype is developed.
engineering verification (EV)	Phase 2 of the development cycle, in which the engineering prototype goes through a design iteration to arrive at a pre-production prototype.

release to manufacturing	The explicit design release from EV to DV, with executive approval. This triggers investments in tooling and other costs.
design verification (DV)	Phase 1 of the manufacturing cycle, in which the contract manufacturer learns to assemble and test the product and develops the manufacturing process, as well as associated manufacturing tests and calibration procedures.
process verification (PV)	Phase 2 of the manufacturing cycle, in which the contract manufacturer tests and hones the process developed during DV and gets ready for mass production.
mass production (MP)	Phase 3 of the manufacturing cycle, in which the ramp-up to mass production begins.
first customer shipment (FCS)	The date when the finished product in final packaging is ready to ship to the first end customers

Figures

About the Author

Elaine Chen is a seasoned engineering and product development executive and consultant with over 20 years of experience bringing products to market in startup environments.

Elaine is the founder and Managing Director of ConceptSpring, a strategic engineering and product development consulting business that helps CEOs and leaders of product teams accelerate the development of disruptive new products and services with hardware and software components. She is also a senior lecturer at MIT, where she introduces entrepreneurship, primary market research, and advanced product development techniques to engineering, MBA and EMBA students.

Elaine is deeply passionate about all aspects of bringing a product to market, from market sizing and analysis through customer research, ideation, user experience, design, development and release to product life cycle management and planned obsolescence. She has brought many hardware and software products to market, in industries spanning industrial automation, robotics, haptics, CAD, consumer electronics, Web, mobile platforms and more. She has built, grown and nurtured several technical organizations from the ground up. She is a co-inventor on 17 patents and has managed IP portfolio strategies for several startups.

Elaine has served as a product management and product development executive for several startups at the VP level, including SensAble Technologies (now part of the Geomagic division of 3D Systems), Zeemote, Zeo, and Rethink Robotics.

Elaine holds a Bachelor of Science and a Master of Science in mechanical engineering from MIT.

Follow Elaine on Twitter: @chenelaine

Subscribe to Elaine's blog on entrepreneurship, product development and leadership: http://blog.conceptspring.com

Resources

Aulet, Bill. *Disciplined Entrepreneurship: 24 Steps to a Successful Startup*. Wiley, 2013.

Blank, Steve. *The Four Steps to the Epiphany*. K&S Ranch 2nd edition, 2013.

Blank, Steve. *The Startup Owner's Manual: The Step-By-Step Guide for Building a Great Company*. K & S Ranch 1st edition, 2012.

Ries, Eric. *The Lean Startup: How Today's Entrepreneurs Use Continuous Innovation to Create Radically Successful Businesses*. Crown Business, 2011.

Made in the USA
Middletown, DE
16 September 2020